DEPARTMENT OF THE ENVIRONMENT

SCOTTISH OFFICE

WELSH OFFICE

ENVIRONMENTAL PROTECTION ACT 1990

WASTE MANAGEMENT

THE DUTY OF CARE

A CODE OF PRACTICE

December 1991

London: HMSO

Applications for reproduction should be made to HMSO
First published 1991
Fourth impression 1992

ISBN 0 11 752557 X

CONTENTS

THE DUTY OF CARE - A CODE OF PRACTICE

INTRODUCTION

i.   This code is issued by the Secretary of State for the Environment, the Secretary of State for Scotland and the Secretary of State for Wales, after consultation, in accordance with section 34(7) of the Environmental Protection Act 1990 ("the Act"). Section 34 of the Act imposes a new duty of care on persons concerned with **controlled waste**. The duty applies to any person who produces, imports, carries, keeps, treats or disposes of controlled waste. Breach of the duty of care is an offence, with a penalty of an unlimited fine if convicted on indictment.

ii.   The purpose of this code is to set out practical guidance for waste holders subject to the duty of care. It recommends a series of steps which would normally be enough to meet the duty. The code cannot cover every contingency; the legal obligation is to comply with the duty of care itself rather than with the code. Annex A gives a detailed explanation of the law on the duty of care.

iii.   The code is divided into:

- step by step advice on following the duty;

- a summary checklist;

- annexes: the law on the duty of care;
  responsibilities under the duty;
  regulations on keeping records;
  an outline of some other legal requirements; and
  a glossary.

STEP BY STEP ADVICE

1. IDENTIFY AND DESCRIBE THE WASTE

1.1.   Anyone subject to the duty of care who has some "controlled waste" should establish what that waste is.

Is it "controlled waste"?

1.2.   First, **is it waste at all?** The term "waste" has a wide meaning, and it is not possible simply to list what is or is not waste, as this can only be determined on the facts of each case. When the question has come before them, the courts have consistently taken the line that waste is defined from the point of view of the person producing or discarding it. With this in mind, where there is any doubt about whether or not something is waste, it may be helpful to ask the following questions *as if from the point of view of the person producing or discarding it:*

> a)   is it what would ordinarily be described as waste?
>
> b)   is it a scrap material?
>
> c)   is it an effluent or other unwanted surplus substance?
>
> d)   does it require to be disposed of as broken, worn out, contaminated or otherwise spoiled?
>
> e)   is it being discarded or dealt with as if it were waste?

1.3.   If the answer to *any* of these questions is "yes", the substance, material or article in question should be considered to be waste. Moreover, anything which is discarded or dealt with as if it were a waste must be presumed to be waste unless it is proved not to be.

1.4. **Is it "controlled waste"?** Waste from households*, commerce or industry is "controlled waste". The main exceptions that are *not* "controlled waste" are waste from agricultural premises, waste from mines and quarries, explosives and most radioactive waste.

What are the problems of the waste?

1.5. Waste cannot be simply divided between the safe and the hazardous. There are safe ways of dealing with any waste, but equally any waste can be hazardous to human health or the environment if it is wrongly managed. Whether waste poses a problem depends on knowing what will happen to it as much as what it is made of. For most waste it is not necessary to know more than what it is in very general terms. But subsequent holders must be provided with a description of the waste that is full enough to enable them to manage the waste properly. Even everyday items may cause problems in handling or treatment.

1.6. In looking for waste problems it may help to ask such questions as:

a) does the waste need a special container to prevent its escape or to keep it from the air? what type of container suits it and what material can the container be made of?

b) can it safely be mixed with any other waste? with what should it not be mixed?

c) can it safely be crushed and transferred from one vehicle to another?

d) can it safely be incinerated? are there special requirements for its incineration such as minimum temperature and combustion time?

e) can it safely be buried in a landfill site with other waste?

---

* Householders are exempt from the duty of care for their own household waste (Annex A, A7).

1.7.   Anything unusual in waste can pose a problem. So can anything out of proportion. Ordinary household waste, and waste from shops or offices, often contains small amounts of potentially hazardous substances. This may not matter if they are mixed in a large quantity of other waste. What are worth identifying as potential problems are significant quantities of an unexpected substance, or unusual amounts of an expected substance, in waste.

1.8.   Note that certain especially dangerous or difficult wastes, "special wastes", are subject to strict legal controls quite apart from and additional to the requirements of the duty of care (Annex D iii).

What goes in a description?

1.9.   A transfer note must be completed, signed and kept by the parties to the transfer if waste is transferred; this is a requirement of the Environmental Protection (Duty of Care) Regulations 1991, and any breach of those regulations is an offence (Annex C Regulations on Keeping Records). This note must state the quantity of waste transferred and how it is packed - whether loose or in a container and, if in a container, the kind of container. There must also be a description of the waste, either separately or combined as a single document with the transfer note. It is good practice to label drums or similar closed containers with a description of the waste. Under the regulations the parties must keep the transfer note and the description for two years. The description should always mention any **special problems**, requirements or knowledge. In addition it should include some combination of:

a)   the type of premises or business from which the waste comes;

b)   the name of the substance or substances;

c)   the process that produced the waste; and

d)   a chemical and physical analysis.

1.10. The description must provide enough information to enable subsequent holders to avoid mismanaging the waste. When writing a description it is open to the holder to ask the manager who will handle the waste what he needs to know. For most wastes, those that need only a simple description, either (a) or (b) will do. Wastes presenting more difficulty will need a fuller description including some or all of (a) to (d). Each element of the description is dealt with in 1.12 to 1.18 below.

Special problems

1.11. The description should always contain any information which might affect the handling of the waste. This should include:

> a) special problems identified under paragraphs 1.5 to 1.7 above;
>
> b) any information, advice or instructions about the handling or disposal of the waste given to the holder from a waste regulation authority* or the suppliers of material or equipment;
>
> c) details of problems previously encountered with the waste;
>
> d) changes to the description since a previous load.

A)   Source of the waste: type of premises or business

1.12. It may sometimes be enough to describe the source of the waste by referring either to the use of the premises where the waste is produced or to the occupation of the waste producer.

---

*   In the case of processes regulated under Integrated Pollution Control in England and Wales, HM Inspectorate of Pollution may give advice to the producer of waste that is relevant to its subsequent management prior to its final disposal or treatment. In Scotland the Integrated Pollution Control authority will be either the chief inspector of HM Industrial Pollution Inspectorate or the River Purification Authority. This advice should also be included in the description.

1.13. Such a "source of waste" description is recommended as the common sense simple description where businesses produce a mixture of wastes none of which has special handling or disposal requirements or where there are no special handling or disposal requirements which cannot be identified from such a simple description. Such a description must make it clear what type of wastes are produced and the contents of the wastes and their proportions must be only such as might be expected.

B) Name of the substance

1.14. The waste may be described by saying what it is made of. This may be in physical and chemical terms or by the common name of the waste where this is equally helpful. Such a description by name is recommended for waste composed of a single simple material.

C) Process producing the waste

1.15. The waste may be described by saying how it was produced. Such a description would include details of materials used or processed, the equipment used and the treatment and changes that produced the waste. If necessary this would include information obtained from the supplier of the materials and equipment.

1.16. This should form part of the description for most industrial wastes and some commercial wastes.

D) Chemical and physical analysis

1.17. A description based on process producing the waste (C) will not go far enough where the holder does not know enough about the source of the waste. It will often not be adequate where:

    a)   wastes, especially industrial wastes, from different activities or processes are mixed; or

b) the activity or process alters the properties or composition of the materials put in.

1.18. For such wastes an analysis will usually be needed; in cases of doubt, the holder may find it helpful to consult the intended waste manager as to whether he needs an analysis to manage the waste properly. Where it is necessary the holder should detail the physical and chemical composition of the waste itself including, where different substances are mixed, their dilutions or proportions. The holder might either provide this information himself or obtain a physical or chemical analysis from a laboratory or from a waste management contractor.

2. KEEP THE WASTE SAFELY

The problem

2.1.   All waste holders must act to keep waste safe against:

a)   corrosion or wear of waste containers;

b)   accidental spilling or leaking;

c)   accident or weather breaking contained waste open and allowing it to escape;

d)   waste blowing away or falling while stored or transported;

e)   scavenging of waste by vandals, thieves, children, trespassers or animals.

2.2.   Holders should protect waste against these while they have it themselves. They should also protect it for its future handling requirements. Waste should reach not only its next holder but a licensed facility or other final destination without escape. Where waste is to be mixed immediately, for example in a transfer station, a civic amenity site or a municipal collection vehicle, it only needs to be packed well enough to reach that immediate destination; preventing its escape after that stage would be up to the next holder. However, there are wastes that may need to reach a disposal or treatment site in their original containers, for example drummed waste. In such cases holders will need to know through how many subsequent hands, under what conditions, for how long and to what ultimate treatment their waste will go in order to satisfy themselves that it is packed securely enough to reach its final destination intact. If an intermediate holder alters waste in any way, by mixing, treating or repacking it, then he will be responsible for observing all this advice on keeping waste safe.

Storing waste securely

2.3.   Security precautions at sites where waste is stored should prevent theft, vandalism or scavenging of waste. Holders should take particular care to secure waste material attractive to scavengers, for example building and demolition materials.

2.4.   Segregration of different categories of waste where they are produced may be necessary to prevent the mixing of incompatible wastes, for example avoiding reactions in mixtures, and may assist the marketing or disposal of waste to specialist outlets. Where segregation is practised on sites, the waste holder should ensure that his employees and anyone else handling waste there are aware of the locations and uses of each segregrated waste container.

Containers

2.5.   Waste handed over to another person should be in some sort of container, which might include a skip. The only reasonable exception would be loose material loaded into a vehicle and then covered sufficiently to prevent escape before being moved. Waste containers should suit the material put in them.

2.6.   It is good practice to label drums or similar closed containers with a note of the contents when stored or handed over; this could be a copy of the waste description.

Waste left for collection

2.7.   Waste left for collection outside premises should be in containers that are strong and secure enough to resist not only wind and rain but also animal disturbance, especially for food waste. All containers left outside for collection will therefore need to be secured or sealed (for example drums with lids, bags tied up, skips covered). To minimise the risks, waste should not be left outside for collection longer than necessary.

3. TRANSFER TO THE RIGHT PERSON

3.1.   Waste may only be handed on to authorised persons or to persons for authorised transport purposes. Full lists of these are in subsections (3) and (4) of section 34 of the Act, a copy of which is in Annex A. This section of the code advises on who these persons are and what checks to carry out *before* making an arrangement or contract for transferring waste.

Public waste collection

3.2.   Local authorities collect waste from households and from some commercial premises. They do this either with their own labour or using private contractors who will be registered carriers (Annex Dii). Private persons are exempt from the duty in connection with their own household waste produced on their premises (Annex A, A7). If there is any doubt about whether or not a particular waste can go in the normal collection, the producer should ask the local authority (the borough, district or islands council).

Using a waste carrier

3.3.   A waste holder may transfer waste to someone who transports it - a waste carrier - who may or may not also be a waste manager (Annex Dii). Subject to certain exemptions, anyone carrying waste in the course of their business, or in any other way for profit, must be registered with a waste regulation authority. Each authority's register of carriers is open to public inspection. Holders may therefore for the purpose of the duty of care use these registers as a reference list of carriers who are authorised to transport waste. However, inclusion on an authority's register is *not* a recommendation or guarantee of a carrier's suitability to accept any particular type of waste. The holder should remain alert to any sign that the waste may not be legally dealt with by a carrier.

3.4.   Anyone intending to transfer waste to a carrier will need to check that the carrier is registered or is exempt from registration. A registered carrier's authority for transporting waste is either his certificate of registration or a copy of his certificate of registration *if it was provided by the waste regulation authority*. The certificate or copy certificate will show the date on which the carrier's registration expires. All copy certificates must be numbered and marked to show that they are copies and have been provided by the waste regulation authority. Photocopies are not valid and **do not** provide evidence of the carrier's registration.

3.5. In all cases other than those involving repeated transfers of waste, the holder should ask to see, and should check the details of, the carrier's certificate or copy certificate of registration. In addition, before using any carrier for the first time, the holder should check with the waste regulation authority with which the carrier is registered that his registration is still valid, even if his certificate appears to be current. The waste regulation authority with whom a carrier is registered is the authority where the carrier has his principal place of business; the authority's name and other details are shown on the certificate. The holder should provide the authority with the carrier's name and registration number as shown on the certificate.

3.6. In practice, the only exempt carriers who might take waste from a holder are:

- charities and voluntary organisations;

- waste collection authorities (local authorities) collecting any waste themselves (though an authority's contractors are not exempt);

- British Rail when carrying waste by rail; and

- ship operators where waste is to be disposed of under licence at sea.

For a transitional period, a carrier may be exempt where he applied for registration before 1 April 1992 and a decision on his application has not been made. In all cases other than those involving repeated transfers of waste, the holder should ask the carrier to confirm the type of exemption under which he transports waste. In addition, before using for the first time a carrier who claims to be exempt, the holder should ask the carrier to provide evidence that the exemption which he claims is valid.

Sending waste for disposal or treatment

3.7.   A "waste manager" is anyone who stores waste or who processes it in some intermediate way short of final disposal or who reclaims waste or who disposes of it. All these activities are licensed, by waste disposal authorities, who issue "disposal licences"; from 1993 these will be called "waste regulation authorities" and "waste management licences". (In this code the term "waste management licence" is used to cover either licence.) The licence includes conditions which usually limit the type and quantity of waste that the operator may deal with and the way in which the waste is managed. The conditions will also govern such matters as operating hours and pollution control on site.

3.8.   *Before* choosing a waste manager as the next person to take some waste a holder will need to:

a)   check that the manager does have a licence; and

b)   establish that the licence permits the manager to take the type and quantity of waste involved.

3.9.   A waste holder should check this not merely by asking the waste manager but by examining his licence. The holder in turn should show the manager the description of the waste involved. If the holder doubts whether the licence does cover his particular waste he can ask the waste manager or, if he is still not satisfied, the authority who issued the licence.

3.10.   Some forms of treatment or disposal of waste do not require licences because they have been exempted by regulations, for example baling up waste paper. If a selected waste manager is not licensed because what he is doing is exempt then he should say so and state which type of exemption he comes under. It is not taking enough care for a holder to consign waste to a contractor who states that he is exempt but does not give the grounds. The holder delivering waste to an exempt waste manager should check that the waste is within the scope of the exemption. The exemptions are limited to specific circumstances and types of waste. If in doubt about the exemption of a particular activity the holder may seek advice from the waste regulation authority for the area where the activity takes place. (For more explanation of licensing and exemptions see Annex Di.)

3.11. A holder of waste should make these same checks on licences and exemptions wherever he delivers waste even if he is not the producer. A carrier should always check that the next holder he delivers to is an authorised person and that the description of the waste he carries is within the licence or exemption of any waste manager to whom he delivers, unless he is only providing the transport to a contract directly between the producer and the waste manager. In that case the producer should make all the checks on the waste manager.

3.12. In Scotland only, local authorities may operate their own disposal sites for publicly collected waste. Such sites in some cases also take waste directly from commerce or industry. Such sites are not licensed but are instead subject to terms and conditions in resolutions passed by the authorities which are broadly similar to those in waste management licences. It is necessary to:

> a) check that such sites are operated by the local waste disposal authority; and
>
> b) establish that the resolution permits the authority to take the type and quantity of waste involved.

In England and Wales, waste disposal sites are being taken out of direct local authority operation; once this is complete both public and private disposal sites will be equally subject to licensing.

Checks on repeated transfers

3.13. Full checks on carriers and waste managers do not need to be repeated if transfers of waste are repetitive - the same type of waste from the same origin to the same destination. The obvious example is waste collection from commercial premises.

3.14. For a series of identical loads making up one transaction there is no need to see the licence of a waste manager or, in Scotland, the local authority's resolution, or the registration certificate of a waste carrier every time a load is handed over. However, licences, registration or evidence of exemption *should* be examined afresh in the following cases:

a) whenever a new transaction is involved, that is if the description or destination of the waste has changed;

b) where several different carriers or disposers are collecting waste at one place and there might be a danger of an unauthorised carrier collecting a load - for example, a construction or demolition site from which several hauliers are taking waste away;

c) as a minimum precaution, the licence, resolution, registration or evidence of exemption should be seen and checked at least once a year even if nothing has changed in a series of repeated transfers;

d) where there has been a change in the licence conditions of the destination.

4. RECEIVING WASTE

4.1. The previous three sections of this code look at transfers from the point of view of the person transferring the waste to someone else. This section offers advice to persons receiving waste, whether at its ultimate destination or as an intermediate holder.

Checking on the source of waste

4.2. Checking is not only in one direction. No-one should accept waste from a source that seems to be in breach of the duty of care. Waste may only come *either* from the person who first produces or imports it *or* from someone who has received it, who must therefore be one of the persons entitled to receive waste (section 3). On the handover of waste, the previous holder and the recipient will complete a transfer note in which the previous holder will declare which category of person entitled to hold waste he is. The recipient should ensure that this is properly completed before accepting waste. Checking back in this way need not be as thorough as checking forward. Recipients should not need to see any waste management licence of the previous holder. However, the first time a carrier delivers waste, it would be reasonable to see either the registration certificate or official copy certificate, or to request confirmation of the type of exemption under which the carrier is transporting the waste.

4.3 Before receiving any waste, a holder should establish that it is contained in a manner suitable for its subsequent handling, and final disposal or reclamation. Where the recipient provides containers he should advise what waste may be placed in them.

4.4 The recipient should also look at the description and seek more information from the previous holder if this is necessary to manage the waste.

Co-operation with the previous holder

4.5.   Anyone receiving waste should co-operate with the previous holder in any steps they are taking to comply with the duty. That means in particular supplying correct and adequate information that the previous holder may *need*.

4.6.   The previous holder needs to know enough about the later handling of his waste - how it is likely to be carried, stored and treated - to pack and describe it properly. The recipient should give such information.

4.7.   Under the regulations on keeping records (Annex C) anyone receiving waste must receive a description and complete a transfer note. The recipient must declare on the transfer note which category of authorised person he is, with details. Before making any arrangement to receive waste a waste manager should show to the previous holder his waste management licence or a statement of the type of exemption from licensing under which he is operating; and a carrier should show his certificate of registration, an official copy of his certificate of registration or evidence of the type of exemption from registration under which he transports waste. It would be sensible for every waste management site office to hold a copy of the waste management licence or a statement of the type of exemption; and for every vehicle used by a carrier to carry an official copy of the carrier's certificate of registration or evidence of the type of exemption under which he transports waste.

5. CHECKING UP

5.1.   The previous sections describe normal procedures for transferring waste from the producer to its final destination. Most of the checking that is reasonable is already built into these procedures and the transfer note system (Annex C). This section gives advice on what further checks are advisable and the action to take when checks show that something is wrong.

Checks after transfer

5.2.   Most waste transfers require no further action from the person transferring waste after the waste has been transferred. A producer is under no specific duty to audit his waste's final destination, however undertaking such an audit and periodic site visits thereafter would be a prudent means of protecting his position by being able to demonstrate the steps he had taken to prevent subsequent illegal treatment of his waste.

5.3.   One exception is where a holder makes arrangements with more than one party, for example a producer arranges two contracts, one for disposal and another for transport to the disposal site. In that case the holder should establish that he not only handed the waste to the carrier but that it reached the disposer.

Checks after receipt

5.4.   Any waste holder, but especially a waste manager receiving waste, has a strong practical interest in the description being correct and containing adequate information. A waste management licence will control the quantities and types of waste that may be accepted and how they may be managed. It is within his capacity as a waste manager to ensure that descriptions of waste received are indeed correct. Anyone receiving waste should make at least a quick visual check that it appears to match the description. For a waste manager it would be good practice to go beyond this by fully checking the composition of samples of waste received.

Causes for concern

5.5.   Every waste holder should be alert for any evidence that suggests that the duty of care is not being observed or that illegal waste handling is taking place. Obvious causes for concern that any holder should notice when he accepts or transfers waste include:

- a)   waste that is wrongly or inadequately described being delivered at a waste management site;

- b)   waste being delivered or taken away without proper packing so that it is likely to escape;

- c)   failure of the person delivering or taking waste to complete a transfer note properly, or an apparent falsehood on the transfer note;

- d)   an unsupported claim of exemption from licensing or registration; and

- e)   failure of waste consigned via a carrier to arrive at a destination with whom the transferring holder has an arrangement.

5.6.   Other causes of concern may come to light. Waste holders do not *have* to check up after waste is handed on, but they may become aware of where it is going, and should act on such information if it suggests illegal or careless waste management.

Action to take with other holders

5.7.   A holder who only suspects that his waste is not being dealt with properly should first of all check his facts, in the first place with the next (or sometimes the previous) holder. This may involve asking that further details should be added to a waste description, for more information about the exact status of the holder for the purposes of waste management licensing or carrier registration, why they may be entitled to an exemption, or simply where waste went to or came from.

5.8.   If a holder is not satisfied with the information or is certain that waste he handles is being wrongly managed by another person then his first action should normally be to refuse to transfer or accept further consignments of the waste in question to or from the person acting wrongly, unless and until the problem is remedied. Such a stop may not be practicable in all cases, for example to avoid breach of a contract to deliver or accept waste or because there is no other outlet immediately available for the waste. Steps should be taken to minimise such inflexibility. One possible measure would be for new waste contracts to provide for termination if a breach of the duty occurs and is not rectified.

5.9.   If an arrangement that has proved wrong in some respect does have to continue temporarily for these reasons then the holder should take stringent precautions. For example where waste has been misdescribed he should analyse further consignments, where waste has been collected or delivered without being properly packed he should inspect each further load and where it has not reached a legitimate destination he should check that each load does arrive in future.

Reporting to waste regulation authorities

5.10.   Waste regulation authorities are responsible for the registration of waste carriers and the licensing of waste management. As such they have a major interest in breaches of the duty of care which might contribute to illegal waste management. They are also equipped with the powers and expertise to prevent or pursue offences and to advise on the legal and environmentally sound management of waste.

5.11.   Holders should tell their waste regulation authority where they know or suspect that:

a)   there is a breach of the duty of care; or

b)   waste is carried by an unregistered carrier not entitled to exemption; or

c)   waste is treated, stored or disposed of without a licence or in a way not permitted by the licence or, in Scotland, by the terms and conditions of an authority's resolution.

6. EXPERT HELP AND ADVICE

6.1. A waste holder may not always have the knowledge or expertise to discharge his duty of care. He should then seek expert help and advice. The holder is still the person responsible for discharging his own duty of care, a responsibility that cannot be transferred to an expert adviser.

Help with analysing, describing and handling waste

6.2. Waste consultants and waste managers offer services to examine waste problems, identify wastes and recommend storage and handling methods. A waste holder may also use an analytical laboratory to establish the nature of unknown waste that he needs to describe. A consultant or laboratory acting only in this way can only advise, and cannot take over the duty of care from the holder. However reliable the expert, the holder himself needs to ask the right questions. Where he is faced with an unknown substance or a known waste but no disposal outlet, he needs to establish not only what it is but what special needs it has for storage, transport or treatment, and what the possible outlets for it may be, whether disposal or reclamation.

Help with choosing a destination

6.3. A consultant may put a waste holder in touch with a legitimate outlet for his waste. Alternatively waste managers or carriers themselves may offer expert advice on a destination for a waste. A large waste management organisation with a variety of disposal or reclamation methods under its control may be able to provide all the waste management services a holder needs. Otherwise it may be necessary to shop around several firms for advice on whether or not they would accept a waste. If a consultant arranges a waste transfer to such an extent that he controls what happens to the waste, he is a broker, and shares responsibility with the two holders directly involved for the proper transfer of the waste.

6.4.   If the next holder is himself a source of advice he may sometimes undertake the analysis and write the waste description on behalf of the previous holder; what he may not do is undertake the checks or complete a transfer note for which the previous holder is responsible.

Help from local authorities

6.5.   A *waste collection authority* will be able to tell a waste holder whether a particular waste may or will be collected as part of their normal public waste collection.

6.6.   *Waste regulation authorities* have expertise in the management of all types of controlled waste but are *not* in a position to offer advice to every waste holder on how he should deal with his waste. Information that **is** held by waste regulation authorities and is available for any enquirer to examine on request is:

a)   the register of waste carriers maintained by each authority; the register is a list of carriers currently registered with that authority; registered carriers are authorised for the purpose of the duty of care to transport controlled waste; and

b)   the public register of waste management licences and, in Scotland only, of resolutions made by authorities, which includes a full copy of each licence issued in the area of the authority, from which a waste holder can check on the waste that a site is entitled to take.

Waste management papers

6.7   Detailed advice on good practice is given in the series of Waste Management Papers issued by the Department of the Environment and published by HMSO. These cover subjects including the handling, treatment and disposal of particular wastes and waste management licensing.

SUMMARY

This section draws together in one place a simple checklist of the main steps that are normally necessary to meet the duty of care. As with the code as a whole, *this does not mean that completing the steps listed here is all that needs to be done under the duty of care.* The checklist cross-refers to key sections of the code for fuller advice.

refer to paragraphs

a) Is what you have waste? if yes,   1.2-1.3

b) is it controlled waste? if yes,   1.4

c) while you have it, protect and store it properly,   2.1-2.4

d) write a proper description of the waste, covering: 1.9-1.10

- any problems it poses;   1.5-1.7

and, *as necessary to others who might deal with it later,* one or more of:

- the type of premises the waste comes from;   1.12-1.13

- what the waste is called;   1.14

- the process that produced the waste; and   1.15-1.16

- a full analysis;   1.17-1.18

e) select someone else to take the waste; they must be one or more of the following and must *prove* that they are:

- a registered waste carrier;   3.3-3.5

- exempt from registration;   3.6

CHECKLIST

-	a waste manager licensed to accept the waste;	3.7-3.9
-	exempt from waste licensing;	3.10
-	a waste collection authority; or	3.2
-	a waste disposal authority operating within the terms of a resolution (Scotland only);	3.12

f) pack the waste safely when transferring it;   2.5-2.7

g) check the next person's credentials   3.4-3.6
when transferring waste to them;   & 3.8-3.14

h) complete and sign a transfer note;   Annex C, C3-C4

i) hand over the description and complete a   1.9-1.10 &
transfer note when transferring the waste;   Annex C, C3-C4

j) keep a copy of the transfer note signed by the
person the waste was given to, and a copy   Annex C,
of the description, for two years;   C5

k) when *receiving* waste, check that the person who
hands it over is one of those listed in (e), or the
producer of the waste, obtain a description from
them, complete a transfer note and keep the   4.2-4.7
documents for two years.   & 5.4

l) Whether transferring *or* receiving waste, be
alert for any evidence or suspicion that the
waste you handle is being dealt with illegally
at any stage, in case of doubt question the
person involved and if not satisfied, alert
the waste regulation authority.   5.5 - 5.11

ANNEX A   THE LAW ON THE DUTY OF CARE

What the duty requires

A.1.   The duty of care is set out in section 34 of the Act (a copy is printed at the end of this Annex). Those subject to the duty must try to achieve the following four things:

> a)   to prevent any other person committing the offences of disposing of "controlled waste" (see Glossary) or treating it, or storing it
>
> - without a waste management licence; or
>
> - breaking the conditions of a licence;or
>
> - in a manner likely to cause pollution or harm to health;
>
> b)   to prevent the escape of waste, that is, to contain it;
>
> c)   to ensure that, if the waste is transferred, it goes only to an "authorised person" or to a person for "authorised transport purposes" - these are listed in subsections (3) and (4) of section 34 of the Act (a copy is printed at the end of this Annex);
>
> d)   when waste is transferred, to make sure that there is also transferred a written description of the waste, a description good enough to enable each person receiving it
>
> - to avoid committing any of the offences under (a) above; and
>
> - to comply with the duty at (b) to prevent the escape of waste.

A.2.   Those subject to the duty must also comply with regulations which require them to keep records and make them available to authorities. These regulations are additional to the code of practice, and they are summarised at Annex C.

A.3.   **Failing to observe the duty or the regulations is a criminal offence.**

Who is under the duty?

A.4.   The duty and therefore this code apply to any

- importer, producer or carrier of controlled waste;

- person who keeps, treats or disposes of controlled waste (a "waste manager")

- broker who has control of controlled waste.

In the code and these Annexes all these persons who are subject to the duty are referred to as "holders" of waste.

A.5.   There is no definition of "*broker*" in the Act. A broker is a person who arranges the transfer of waste which he does not himself hold to such an extent that he controls what happens to it.

A.6.   Employers are responsible for the acts and omissions of their employees. They therefore should provide adequate equipment, training and supervision to ensure that their employees observe the duty of care.

Exemption for householders

A.7.   The only exception to the duty is for occupiers of domestic property for "household waste" (see Glossary) from the property. Note that this does *not* exempt:

a)   a householder disposing of waste that is not from his property (for example waste from his workplace; or waste from his neighbour's property); or

b)   someone who is not the occupier of the property (for example a builder carrying out works to a house he does not occupy is subject to the duty for the waste he produces).

What each waste holder has to do

A.8.   A waste holder is not absolutely responsible for ensuring that all the aims of the duty of care (listed in A.1.) are fulfilled. A holder is only expected to take measures that are

a)   reasonable in the circumstances, *and*

b)   applicable to him in his capacity.

A.9.   The *circumstances* that affect what is reasonable will include:

a)   what the waste is;

b)   the dangers it presents in handling and treatment;

c)   how it is dealt with; and

d)   what the holder might reasonably be expected to know or foresee.

A.10.   The *capacity* of the holder is who he is, how much control he has over what happens to the waste and in particular what his connection with the waste is; is he the importer, producer, carrier, keeper, treater, disposer or broker? Different measures will be reasonable for each.

A.11.   Waste holders can be responsible only for waste which is at some time under their control; it is not a breach of the duty to fail to take steps to prevent someone else from mishandling any other waste. However, a holder's responsibility for waste which he at any stage controls extends to what happens to it at other times, insofar as he knows or might reasonably foresee. Further guidance on the reasonable limits of responsibilities is in Annex B.

Offences and penalties

A.12. Breach of the duty of care is a criminal offence. It is an offence irrespective of whether or not there has been any other breach of the law or any consequent harm or pollution. The offence is punishable by a fine of up to £2,000 on summary conviction or an unlimited fine on conviction on indictment.

The code of practice

A.13. This code of practice has statutory standing. The Secretary of State is obliged by the Act to issue practical guidance on how to discharge the duty; this code forms that guidance. The code is admissible as evidence in court. The court shall then take the code into account in determining any question to which it appears relevant. The code may therefore be used by a court in deciding whether or not an accused has complied with his duty of care.

c.43 *Environmental Protection Act 1990*

Duty of care etc. as respects waste

Duty of care etc. as respects waste. **34.**-(1) Subject to subsection (2) below, it shall be the duty of any person who imports, produces, carries, keeps, treats or disposes of controlled waste or, as a broker, has control of such waste, to take all such measures applicable to him in that capacity as are reasonable in the circumstances -

> (a) to prevent any contravention by any other person of section 33 above;
>
> (b) to prevent the escape of waste from his control or that of any other person; and
>
> (c) on the transfer of the waste to secure -
>
>> (i) that the transfer is only to an authorised person or to a person for authorised transport purposes; and
>>
>> (ii) that there is transferred such a written description of the waste as will enable other persons to avoid a contravention of that section and to comply with the duty under this subsection as respects the escape of waste.

(2) The duty imposed by subsection (1) above does not apply to an occupier of domestic property as respects the household waste produced on the property.

(3) The following are authorised persons for the purpose of subsection (1)(c) above -

> (a) any authority which is a waste collection authority for the purposes of this Part;
>
> (b) any person who is the holder of a waste management licence under section 35 below or of a disposal licence under section 5 of the Control of Pollution Act 1974;

1974 c. 40.

> (c) any person to whom section 33(1) above does not apply by virtue of regulations under subsection (3) of that section;
>
> (d) any person registered as a carrier of controlled waste under section 2 of the Control of Pollution (Amendment) Act 1989;

1989 c. 14.

(e) any person who is not required to be so registered by virtue of regulations under section 1(3) of that Act; and

(f) a waste disposal authority in Scotland

(4) The following are authorised transport purposes for the purposes of subsection (1)(c) above -

(a) the transport of controlled waste within the same premises between different places in those premises;

(b) the transport to a place in Great Britain of controlled waste which has been brought from a country or territory outside Great Britain not having been landed in Great Britain until it arrives at that place; and

(c) the transport by air or sea of controlled waste from a place in Great Britain to a place outside Great Britain;

and "transport" has the same meaning in this subsection as in the Control of Pollution (Amendment) Act 1989.

(5) The Secretary of State may, by regulations, make provision imposing requirements on any person who is subject to the duty imposed by subsection (1) above as respects the making and retention of documents and the furnishing of documents or copies of documents.

(6) Any person who fails to comply with the duty imposed by subsection (1) above or with any requirement imposed under subsection (5) above shall be liable -

(a) on summary conviction, to a fine not exceeding the statutory maximum; and

(b) on conviction on indictment, to a fine.

(7) The Secretary of State shall, after consultation with such persons or bodies as appear to him representative of the interests concerned, prepare and issue a code of practice for the purpose of providing to persons practical guidance on how to discharge the duty imposed on them by subsection (1) above.

(8) The Secretary of State may from time to time revise a code of practice issued under subsection (7) above by revoking, amending or adding to the provisions of the code.

(9) The code of practice prepared in pursuance of subsection (7) above shall be laid before both Houses of Parliament.

(10)   A code of practice issued under subsection (7) above shall be admissable in evidence and if any provision of such a code appears to the court to be relevant to any question arising in the proceedings it shall be taken into account in determining that question.

(11)   Different codes of practice may be prepared and issued under subsection (7) above for different areas.

ANNEX B RESPONSIBILITIES UNDER THE DUTY OF CARE

B.1.   The main sections of the code address all waste holders who are subject to the duty of care. In law all the responsibilities for a waste under the duty are spread among all those who hold that waste at any stage, but responsibilities are not spread evenly. Some holders will have greater or less responsibility for some aspects of the duty, according to their connection with the waste (Annex A9-A11 above). This Annex offers advice for each category of waste holder on their particular responsibilities. This Annex is neither comprehensive nor self-contained guidance for particular categories of waste holder. It only draws out questions of the allocation of responsibility. All waste holders should follow the main advice in the code of practice.

Waste producers

B.2.   The duty of care applies to a person who "produces" waste. No definition of "produces" is offered in the Act. The starting point for deciding who is a producer of any waste is deciding how it has become waste. There is a definition of "waste" in section 75 of the Act, continuing the effect of the definition in the Control of Pollution Act 1974, and this can be interpreted with the help of past court judgements. Something may become waste either by being changed in some way (for example by being contaminated or broken) or, if there is no such change, by a decision or change of attitude on the part of the holder of the object or substance (deciding that it is surplus or unwanted). Where an action is involved the waste producer will be the person committing the action. Where no action is involved the waste producer will be the person holding the object or substance who takes the decision that it is waste. In most cases this should be quite unambiguous, but the circumstances of contracting, and especially the construction and demolition industries, have given rise to conflicting views of who is a waste producer and who is subject to the duty of care.

B.3.   In such cases the producer of construction or demolition waste may be regarded as the person undertaking the works which give rise to that waste, not the person who issues instructions or lets contracts which give rise to waste. The client for works, although he may take decisions as a result of which waste is created, is not himself producing the waste created by the works. Where there are several contractors and sub-contractors on site, the producer of a particular waste is the particular contractor or sub-contractor who (or whose employees) takes an action which creates waste, or, who begins to treat something as if it were waste (by discarding it). Where a client or contractor makes arrangements for the carriage or disposal of waste, for example by letting a disposal sub-contract to a haulier for waste produced on site by a demolition sub-contractor, then that client or contractor will be acting as a broker in respect of the transfer between the two sub-contractors; in such a case all three parties will be under the duty. In practice it is likely that every contractor involved on a site will either be producing or carrying away some waste and will be subject to the duty as producer or carrier and therefore liable to account for the measures they have taken to comply with the duty in respect of that waste. It would also be prudent for the client to take reasonable steps to ensure that all the contractors he employs or supervises comply with the duty of care.

B.4.   Waste producers are solely responsible for the care of their waste while they hold it. Waste producers are normally best placed to know what their waste is and to choose the disposal or treatment method, if necessary with expert help and advice. They bear the main responsibility for the description of waste which leaves them being correct and containing all the information necessary for handling, treatment or disposal. If they also select a final treatment or disposal destination then they share with the waste manager of that destination responsibility for ensuring that the waste falls within the terms of any licence or exemption.

B.5.   Producers bear the main responsibility for packing waste to prevent its escape in transit. Waste leaving producers should be packed in a way that subsequent holders could rely on.

B.6.   Using a registered or exempt carrier does not necessarily let a producer out of all responsibility for checking the later stages of the disposal of his waste. The producer and the disposer may sometimes make all the arrangements for the disposal of waste, and then contract with a carrier simply to convey the waste from one to the other. Such a case is very little different in practice from that where there is no intermediate carrier involved. If a producer arranges disposal then he should undertake the same care in selecting the disposer as if he were delivering the waste himself, however many other parties there are between.

B.7.   It is not possible to draw a line at the gate of producers' premises and say that their responsibility for waste ends there. A producer is responsible according to what he knows or should have foreseen. So if he hands waste to a carrier not only should it be properly packed when transferred, but the producer should take account of anything he sees or learns about the way in which the carrier is subsequently handling it. The producer would not be expected to follow the carrier but he should be able to see whether the waste is loaded securely for transport when it leaves, and he may come to learn or suspect that it is not ending up at a legitimate destination. A producer may notice a carrier's lorries returning empty for further loads in a shorter time than they could possibly have taken to reach and return from the nearest lawful disposal site; or a producer may notice his carrier apparently engaged in the unlawful dumping of someone else's waste. These would be grounds for suspecting illegal disposal of his own waste. The same reasoning applies when a producer makes arrangements with a waste manager for the treatment or disposal of waste. The producer shares the blame for illegal treatment of his waste if he ignores evidence of mistreatment; a producer should act on knowledge to stop the illegal handling of waste (paragraphs 5.5 to 5.11).

Waste importers

B.8.   Waste importers should act in most respects as if they were the producers of the waste they import. Regulations governing the transfrontier shipment of hazardous waste ensure that the final destination of the waste is already established. The importer bears the main responsibility for ensuring the adequacy of the accompanying description, the packaging of the waste on entry to Great Britain and the fitness of the destination to deal with the waste.

Waste carriers

B.9.   A waste carrier is responsible for the adequacy of packaging while waste is under his control, he should not rely totally on how it is packed or handed over by the previous holder. For any waste he should at least look at how it is contained to ensure that it is not obviously about to escape. His own handling in transferring and transporting waste should take account of how it is packed, or if necessary he may need to repack it or have it repacked to withstand handling.

B.10. A waste carrier would not normally be expected to take particular measures to provide a new description of the waste he carried unless he altered it in some way. The description would normally be provided by the waste producer, unless the producer is a householder not subject to the duty, in which case the person first taking waste from the householder would have to ensure that a description was furnished to the next holder. When accepting any waste a carrier should make at least a quick visual inspection to see that it appears to match the description, but he need not analyse the waste unless there is reason to suspect an anomaly or the waste is to be treated in some way not foreseen by the producer.

B.11. If a carrier or any other intermediate holder does alter waste in any way, by mixing, treating or repacking it, he will need to consider whether a new description is necessary. The description received may continue to serve for waste that is merely compacted or mixed with similar waste, but if it has deteriorated or decomposed or been altered in any way that matters for handling and disposal then a new description should be written.

B.12. Where there is a contract between the producer and the waste manager and a carrier is contracted solely to provide transport, he may rely on the producer checking the scope of the licence or exemption of the waste manager to whom the waste is delivered. In other cases the carrier should check this himself.

Waste brokers

B.13. A waste broker arranging the transfer of waste between a producer and a waste manager, to such an extent that he controls what happens to the waste, is taking responsibility for the legality of the arrangement. He should ensure that he is as well informed about the nature of the waste as if he were discharging the responsibilities of both producer and waste manager. He is as responsible as either for ensuring that a correct and adequate description is transferred, that the waste is within the scope of any waste licence or exemption, that it is carried only by a registered or exempt carrier and that documentation is properly completed. If he does not directly handle the waste then he cannot be held responsible for its packaging, but he should undertake the same level of checks after transfer and the same action on any cause for suspicion as any other waste holder.

Waste managers

B.14. Waste managers, like waste carriers, should normally be able to rely on the description of waste supplied to them. However in treating or disposing of waste they are in a stronger position to notice discrepancies between the description and the waste and therefore bear a greater responsibility for checking descriptions of waste they receive. Sample checks on the composition of waste received would be good practice.

B.15. Waste managers are also in a good position to notice evidence that waste has been wrongly dealt with or falsely documented before it reaches them. For example a waste manager may receive waste which is documented as coming directly from one producer but which shows signs of having been mixed or treated at some intermediate stage. A manager is to some extent responsible for following up evidence of previous misconduct just as he is for subsequent mismanagement of waste, that is to the extent that he knew or should have foreseen it and to the extent that he can control what happens.

ANNEX C   REGULATIONS ON KEEPING RECORDS

C.1.   The Environmental Protection (Duty of Care) Regulations 1991 (made under section 34(5) of the Act) require all those subject to the duty to make records of waste they receive and consign, keep the records and make them available to waste regulation authorities.

C.2.   The regulations require each party to any transfer to keep a copy of the description which is transferred. An individual holder might transfer onward the description of the waste that he received unchanged in which case it would be advisable for the sake of clarity to endorse the description for onward transfer to the effect that the waste was sent onwards as received. If a different description of waste is transferred onwards, whether or not this reflects any change in the nature or composition of the waste, then copies of both descriptions must be made. The holder making the copy need not be the author of the description, which will often be written only by the producer and reused unchanged by each subsequent holder.

C.3.   The regulations also require the parties to complete, sign and keep a transfer note. The transfer note contains information about the waste and about the parties to the transfer.

C.4.   While all transfers of waste must be documented, nothing in the regulations requires *each* individual transfer to be separately documented. In the case of, for example, weekly or daily collections of waste from outside shops or commercial premises, or the removal of a large heap of waste by multiple lorry trips this would be unrealistic because of the number of individual loads and the absence of any necessary contact between the parties to the transfer. It would be reasonable for a single transfer note to cover multiple consignments of waste transferred at the same time or over a period (not exceeding a year) provided that the description and all the other details on the transfer note (who and what the parties are) are the same for all consignments covered by the note. A single transfer note could not cover a series of transfers between different holders nor transfers of wastes of different descriptions.

Duty of Care: Controlled Waste Transfer Note

Section A - Description of Waste

1. Please describe the waste being transferred:

2. How is the waste contained?

   Loose ☐   Sacks ☐   Skip ☐   Drum ☐   Other ☐ → *please describe:*

3. What is the quantity of waste (number of sacks, weight etc):

Section B - Current holder of the waste

1. Full Name (BLOCK CAPITALS):

2. Name and address of Company:

3. Which of the following are you? (Please ✓ one or more boxes)

   producer of the waste ☐

   importer of the waste ☐

   waste collection authority ☐

   waste disposal authority ☐
   (Scotland only)

   holder of waste disposal or ☐ → *Licence number:*
   waste management licence   *Issued by:*

   exempt from requirement to ☐ → *Give reason:*
   have a waste disposal or
   waste management licence

   registered waste carrier ☐ → *Registration number:*
                                  Issued by:

   exempt from ☐ → *Give reason:*
   requirement to register

Section C - Person collecting the waste

1. Full Name (BLOCK CAPITALS):

2. Name and address of Company:

3. Which of the following are you? (Please ✓ one or more boxes)

waste collection authority	☐ →	*holder of waste disposal or waste management licence*	*Licence number:* *Issued by:*
waste disposal authority (Scotland only)	☐ →	*exempt from requirement to have a waste disposal or waste management licence*	*Give reason:*
	☐ →	*registered waste carrier*	*Registration number:* *Issued by:*
exporter	☐ →	*exempt from requirement to register*	*Give reason:*

Section D

1. Address of place of transfer/collection point:

2. Date of transfer:

3. Time(s) of transfer (for multiple consignments, give 'between' dates):

4. Name and address of broker who arranged this waste transfer (if applicable):

5. Signed:

Signed:

Full name:
(BLOCK CAPITALS)
Representing:

Full Name:
(BLOCK CAPITALS)
Representing:

C.5. The regulations require these records (both the descriptions and the transfer notes) to be kept for at least two years. Holders must provide copies of these records if requested by any waste regulation authority.

C.6. One purpose of documentation is to create an information source of use to other holders. It is open to holders to ask each other for details from records, especially to check what happened to waste after it was consigned. A holder might draw conclusions and alert the waste regulation authority to any suspected breach of the duty if such a request were refused.

C.7. There is no compulsory form for keeping these records. It is recognised that a number of holders already keep records of waste in a manner that meets the requirements of the regulations with little or no further adaptation. A suggested standard form for voluntary use is included in this Annex.

C.8. Breach of any provision of the regulations is an offence.

ANNEX D OTHER LEGAL CONTROLS

D.1. This code offers guidance on the discharge of a waste holder's duty of care under section 34 of the Act. Holders are also subject to other statutory requirements, some of the most important of which are set out here.

i. Waste disposal and waste management licensing

Current Controls

D.2. In *England and Wales* under the Control of Pollution Act 1974, *waste disposal licences* are required to authorise the:

> a) **deposit** of controlled waste on land;
>
> b) **disposal** of controlled waste by means of plant or equipment; and
>
> c) **use of plant or equipment** to deal with controlled waste in one of the ways prescribed in Schedule 5 of the Collection and Disposal of Waste Regulations 1988 (SI 1988/819, published by HMSO). Briefly, treating waste by baling, compacting, incinerating, pulverising, sorting, storing, processing, shredding, or composting may require a licence.

D.3. In *England and Wales, exemptions* from the requirement to have a waste disposal licence for any of the activities listed above are to be found in:

> a) Schedule 6 of the Collection and Disposal of Waste Regulations 1988;

b) Regulation 4 of the Control of Pollution (Landed Ships' Waste) Regulations 1987 (SI 1987/402, published by HMSO), with the amendments made by the Control of Pollution (Landed Ships' Waste) (Amendment) Regulations 1989 (SI 1989/65, published by HMSO); and

c) the Disposal of Controlled Waste (Exceptions) Regulations 1991 (SI 1991/508, published by HMSO).

D.4. In *Scotland* the position is broadly similar but licensing requirements and exceptions are set out in the Control of Pollution (Licensing of Waste Disposal)(Scotland) Regulations 1977 (SI 2006 (S153), published by HMSO).

Future Controls

D.5. From April 1993, the new system of *waste management licences* under the Environmental Protection Act 1990 will be introduced. Under the Act, licences will be required to authorise the:

a) deposit of controlled waste on land;

b) treatment, keeping or disposal of controlled waste on land;

c) treatment, keeping or disposal of controlled waste by means of mobile plant; and

d) treatment, keeping or disposal of controlled waste in a manner likely to cause pollution of the environment or harm to human health.

D.6. *Exemptions* from these requirements will be prescribed in new regulations under section 33(3) of the Act which will replace those, currently in force in England and Wales, in the Collection and Disposal of Waste Regulations 1988.

ii. The registration of waste carriers

D.7. Subject to certain provisions, section 1(1) of the Control of Pollution (Amendment) Act 1989 ("the 1989 Act") makes it an offence to transport controlled waste without being registered. The requirement to register applies to any person who transports controlled waste to or from any place in Great Britain in the course of any business of his or otherwise with a view to profit. For this purpose, "transport" includes the transport of waste by road, rail, air, sea or inland waterway. The Controlled Waste (Registration of Carriers and Seizure of Vehicles) Regulations 1991 (SI 1991/1624, published by HMSO) require each waste regulation authority to establish and maintain a register of waste carriers; and set out the basis on which the registration system operates. Guidance on registration is provided in DOE Circular 11/91, Welsh Office Circular 34/91 and Scottish Office Environment Department Circular 18/91.

D.8. Anyone subject to the duty of care must ensure that, if waste is transferred, it is transferred only to an authorised person or to a person for authorised transport purposes. Among those who are "authorised persons" are:

a) any person registered with a waste regulation authority as a carrier of controlled waste; and

b) any person who is exempt from registration by virtue of regulations made under section 1(3) of the 1989 Act. The exemptions at present in force are set out in regulation 2 of the Controlled Waste (Registration of Carriers and Seizure of Vehicles) Regulations 1991.

D.9. It is not an offence under section 1(1) of the 1989 Act to transport controlled waste without being registered in the circumstances set out in (a) - (c) below. The following are also "authorised transport purposes" for the duty of care:

a) the transport of controlled waste between different places within the same premises;

b) the transport to a place in Great Britain of controlled waste which has been brought from a country or territory outside Great Britain and is not landed in Great Britain until it arrives at that place. This means that the requirement to register as a carrier applies only from the point at which imported waste is landed in Great Britain; and

c) the transport by air or sea of controlled waste being exported from Great Britain.

iii. Special waste

D.10. Certain difficult or dangerous wastes ("special wastes") are subject to additional requirements in the Control of Pollution (Special Waste) Regulations 1980 (SI 1980/1709, published by HMSO) made under section 17 of the Control of Pollution Act 1974 . (These will eventually be superseded by regulations under section 62 of the Environmental Protection Act 1990.)

D.11. Special waste is subject to the duty of care, including the advice in the code of practice and the requirements of The Environmental Protection (Duty of Care) Regulations 1991, in the same way as any other controlled waste. Compliance with the duty of care does not in any way discharge the need also to comply with the special waste regulations.

D.12. It should be noted that the completion of special waste consignment note procedures, or of the similar procedures for landed ships' waste*, go some way towards meeting the purposes of the duty of care regulations. The information in these consignment notes will include almost all the information required in a duty of care transfer note, except some of the details relating to authorised persons. These could be added so that no separate note need be made. A copy of a consignment note would be furnished on request to an authority in place of a transfer note.

---

* The Control of Pollution (Landed Ships' Waste) Regulations 1987 (SI 1987/402, published by HMSO).

iv. Road transport of dangerous substances

D.13. Waste holders have obligations in respect of the regulations and associated codes of practice concerned with the transport of dangerous substances. For national transport the relevant regulations are:

a) the Dangerous Substances (Conveyance by Road in Road Tankers and Tank Containers) Regulations 1981 (SI 1981/1059, published by HMSO);

b) the Classifiation, Packaging and Labelling of Dangerous Substances Regulations 1984 (SI 1984/1244, published by HMSO);

c) the Road Traffic (Carriage of Dangerous Substances in Packages etc) Regulations 1986 (SI 1986/1951, published by HMSO), and (Amendment) Regulations of 1989 (SI 1989/105, published by HMSO); and

d) the Dangerous Substances in Harbour Areas Regulations 1987 (SI 1987/37, published by HMSO).

v. International waste transfers

D.14. Holders have obligations in respect of the import or export of waste. At present these obligations are set in the Transfrontier Shipment of Hazardous Waste Regulations 1988 (SI 1988/1562, published by HMSO). Completion of the consignment note used in connection with these regulations would include almost all the information required on a transfer note, and may be used in a similar manner to a special waste consignment note (D.12 above).

vi. Health and safety

D.15. Waste holders also have a duty to ensure, so far as is reasonably practicable, the health and safety of their employees and other persons who may be affected by their actions in connection with the use, handling, storage or transport of waste, by virtue of sections 2 and 3 of the Health and Safety at Work etc Act 1974.

D.16. Holders also have a duty to comply with the Control of Substances Hazardous to Health Regulations 1988 (SI 1988/1657, published by HMSO) (and in particular the requirement to carry out an assessment of the risks of their activities).

ANNEX E   GLOSSARY OF TERMS USED IN THIS CODE OF PRACTICE

The Act: the Environmental Protection Act 1990.

The 1989 Act: the Control of Pollution (Amendment) Act 1989.

Carrier: a person who transports controlled waste, within Great Britain, including journeys into and out of Great Britain.

A **registered carrier** is registered with a waste regulation authority under the 1989 Act.

An **exempt carrier** is a waste carrier who is not required to register under the 1989 Act because:

a)   he is exempt from registration by virtue of regulation 2 of the Controlled Waste (Registration of Carriers and Seizure of Vehicles) Regulations 1991; or

b)   he is transporting waste in the circumstances set out in section 1(2) of the 1989 Act. These circumstances are also "authorised transport purposes" as defined in section 34(4) of the 1990 Act.

Controlled waste: as defined in section 75 of the Act, that is, household, commercial and industrial waste, as modified by the Collection and Disposal of Waste Regulations 1988.

Holder: a person who imports, produces, carries, keeps, treats, or disposes of controlled waste or, as a broker, has control of such waste.

A **licensed waste manager** is one in possession of a licence under section 35 of the Act *or*, for a transitional period, under section 5 of the Control of Pollution Act 1974. Note that not all persons licensed under section 35 are disposers or reclaimers; licences are also issued for the treatment or keeping of waste by intermediate holders.

An **exempt waste manager** is one exempted from licensing by regulations under section 33(3) of the Act.

Producer: a person whose actions give rise to controlled waste including deciding to discard an article, material or substance.

Waste collection authority: local authority responsible for collecting waste; as defined in section 30(3) of the Act.

Waste disposal authority: as defined in section 30(2) of the Act, that is a local authority responsible for arranging the disposal of publicly collected waste. **In Scotland**, a waste disposal authority may itself act as a waste manager.

Waste regulation authority: as defined in section 30(1) of the Act; the authority charged with the issue of waste management licences under section 35 of the Act (or, for a transitional period until section 35 of the Act comes into force, the authority charged with issuing waste disposal licences under section 5 of the Control of Pollution Act 1974).

Printed in the United Kingdom for HMSO
Dd.295808, 3/92, C200, 3390/3, 5673, 192398